I0805581

CHEERS FOR CAREERS!

I WANT TO BE A LANDSCAPER

by Julie Murray

Cody Koala
An Imprint of Pop!
popbooksonline.com

Hello! My name is Cody Koala

This book is filled with videos, puzzles, games, and more! Scan the QR codes* while you read, or visit the website below to make this book pop.

popbooksonline.com/landscaper

*Scanning QR codes requires a web-enabled smart device with a QR code reader app and a camera.

abdobooks.com

Published by Pop!, a division of ABDO, PO Box 398166, Minneapolis, Minnesota 55439. Copyright ©2025 by Abdo Consulting Group, Inc. International copyrights reserved in all countries. No part of this book may be reproduced in any form without written permission from the publisher. Cody Koala™ is a trademark and logo of Pop!.

Printed in the United States of America, North Mankato, Minnesota.

052024
082024

THIS BOOK CONTAINS RECYCLED MATERIALS

Cover Photo: Shutterstock Images
Interior Photos: Shutterstock Images; Getty Images
Editor: Elizabeth Andrews and Grace Hansen
Series Designer: Colleen McLaren

Library of Congress Control Number: 2023947428

Publisher's Cataloging-in-Publication Data

Names: Murray, Julie, author.
Title: I want to be a landscaper / by Julie Murray
Description: Minneapolis, Minnesota : Pop!, 2025 | Series: Cheers for careers! | Includes online resources and index
Identifiers: ISBN 9781098246051 (lib. bdg.) | ISBN 9781098246617 (ebook)
Subjects: LCSH: Landscaping industry--Juvenile literature. | Gardening--Juvenile literature. | Landscape gardening--Juvenile literature. | Servicing trades--Juvenile literature. | Occupations--Juvenile literature.
Classification: DDC 635.906--dc23

Table of Contents

Chapter 1

Outdoor Spaces

Landscapers design and **install** things in outdoor spaces. They work with grass, flowers, shrubs, bushes, and trees. They also work with stone and lighting. They make places beautiful!

Landscapers can work inside malls or buildings that have green spaces.

Watch a video here!

Chapter 2

Skills Needed

Being a landscaper includes physical work. There is a lot of bending and lifting. Sometimes landscapers have to move heavy objects. They use power tools and operate machinery.

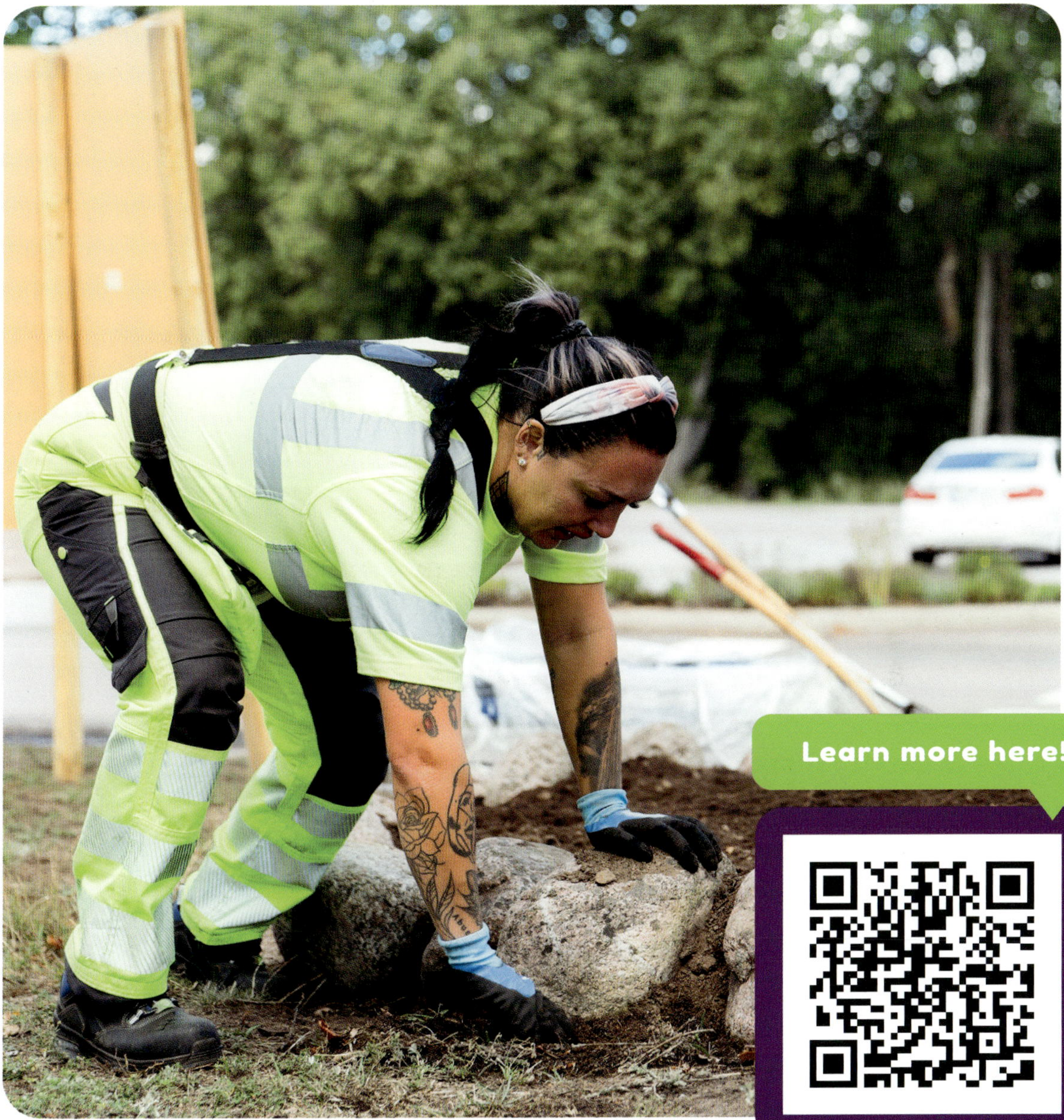

Learn more here!

Landscapers must be **organized**. They create plans called blueprints that show all parts of a project. Blueprints must include plants, stonework, and trees. They can also include features such as lighting and water fountains.

Landscapers must be good communicators. They work with **clients** such as city planners and home

and business owners. Communication between a landscaper and client helps projects run smoothly.

Chapter 3

How to Become a Landscaper

Experience is needed to become a landscaper. An **apprenticeship** is the best way to get hands-on experience. Many companies offer these opportunities.

Explore links here!

Some people get a **degree** in **landscape architecture** or **horticulture**. These are programs that study specific areas of landscaping.

Knowledge of plants and trees is an important part of a landscaper's job!

Chapter 4

Different Jobs

Some landscapers design spaces by creating plans. They take careful measurements of all the areas. They order the needed plants and materials.

Complete an activity here!

hose and nozzle
rake
shears
shovels
cultivator
mower
wheelbarrow

Some landscapers help **install** the products. They plant seeds and plants or lay stone materials. Others work with sprinkler systems. Some care for the grass and mulch areas. Many tools are needed in landscaping.

Being a landscaper can be a rewarding career. Working to create beautiful green spaces is a great job!

Working with plants can have positive effects on people.

Making Connections

Text-to-Self

Would you like to be a landscaper? Why or why not?

Text-to-Text

Is there another book you have read that shows an area that has been landscaped?

Text-to-World

Can you think of a place you have visited or seen a picture of that has beautiful landscaping? Where was it? What did it look like?

Glossary

apprenticeship – an arrangement in which someone learns a trade under an expert.

client – a person who hires or uses the services of a professional of some type.

degree – a title given by a university or college to a student who has completed a program of study.

horticulture – the science and art of growing plants.

install – to put into position and make ready for use.

landscape architecture – the art of planning and designing outdoor areas.

organized – able to plan things carefully and keep things neat.

Index

Online Resources

popbooksonline.com

Thanks for reading this Cody Koala book!

This book is filled with videos, puzzles, games, and more! Scan the QR codes* while you read, or visit the website below to make this book pop.

popbooksonline.com/landscaper

*Scanning QR codes requires a web-enabled smart device with a QR code reader app and a camera.